I0494214

Disclaimer

Book Title: Proceedings of the GSA 3D Imaging Workshop, February 28, 2011 (NIST TN 1699)

Book Author: Geraldine S. Cheok; Marek Franaszek; Kamel S. Saidi;

Book Abstract: The General Services Administration (GSA) has been involved in 3D imaging since the early 2000s. Since 3D imaging was a relatively new technology at that time, GSA needed to develop and provide initial guidance to their project managers (PMs) on 3D imaging technology and on how to develop the scope and specifications (scope of work or SOW) for 3D imaging projects. In the mid to late 2000s, GSA developed initial guidance, in the form of a template for the SOW, based on the following criteria: ease of use for GSA project managers and the need to accommodate widely varying project types and project requirements. This report presents a summary of the workshop discussions. A list of the workshop participants is given in Appendix A. The agenda is given in Appendix B. A list of the questions that was sent out to the workshop participants prior to the workshop is given in Appendix C. Comments on the GSA's SOW for 3D-4D-BIM services received from the contractors are reproduced in Appendix D.

Citation: NIST TN - 1699

NIST TN 1699

Proceedings of the General Services Administration 3D Imaging Workshop, February 28, 2011

Geraldine S. Cheok
Marek Franaszek
Kamel S. Saidi

National Institute of
Standards and Technology
U.S. Department of Commerce

NIST TN 1699

Proceedings of the General Services Administration 3D Imaging Workshop, February 28, 2011

June, 2011

Geraldine S. Cheok, Engineering Laboratory
Marek Franaszek, Engineering Laboratory
Kamel S. Saidi, Engineering Laboratory

U. S. Department of Commerce
Gary Locke, Secretary

National Institute of Standards and Technology
Patrick D. Gallagher, Director

Disclaimer

Table of Contents

Page left intentionally blank.

1 Introduction

The General Services Administration (GSA) has been involved in 3D imaging since the early 2000s. Since 3D imaging was a relatively new technology at that time, GSA needed to develop and provide initial guidance to their project managers (PMs) on 3D imaging technology and on how to develop the scope and specifications (scope of work or SOW) for 3D imaging projects. In the mid to late 2000s, GSA developed initial guidance, in the form of a template for the SOW, based on the following criteria: ease of use for GSA project managers and the need to accommodate widely varying project types and project requirements.

GSA has had several 3D imaging projects where the SOW template was used and has received some feedback on needed improvements to the SOW. To determine what further revisions are needed in the SOW template, GSA conducted a one-day workshop at their Washington D.C. headquarters on February 28, 2011 for their 3D imaging Indefinite Delivery Indefinite Quantity (IDIQ) contractors. There were 18 participants who attended the workshop in person with 20 participants who participated remotely. The workshop was facilitated by the National Institute of Standards and Technology (NIST). The objective of the workshop was to solicit information from the IDIQ contractors to improve how GSA develops the SOW for 3D imaging projects. GSA will use this information to revise the SOW template and the BIM (Building Information Model) Guide Series 03.

This report presents a summary of the workshop discussions. A list of the workshop participants is given in Appendix A. The agenda is given in Appendix B. A list of the questions that was sent out to the workshop participants prior to the workshop is given in Appendix C. Comments on the GSA's SOW for 3D-4D-BIM services received from the contractors are reproduced in Appendix D.

2 Pre-workshop Survey

In preparation for the workshop, a list of questions (Appendix C) was sent to eight (8) IDIQ contractors. Some IDIQ contractors did not respond to all the questions. This section presents a summary of the results from the survey questions.

Note that the green text in this report represents comments received from the workshop participants after the workshop was conducted. That is, a draft of the workshop proceedings was sent to the participants, and they were asked to comment on the draft. Their comments to this draft are included in this report and are represented by green text. The green text does not include any comments by GSA or NIST.

> **Question 1A**: GSA has identified several use-cases for laser scanning: Exterior, Interior – offices, Interior – mechanical/utility, Historic, Sculpture
>
> Are these the main ones? **7 – Yes,   1 - No**
>
> *Comments*:
> *Additional response received after workshop: Yes*

Question 1B: Are there any missing cases? **7 – Yes,   0 - No**

Comments:

1. *Verification of as-built location of items*
2. *Used to satisfy construction as-built and phased as-built requirements*
3. *Scanning assemblies prior to close-up for lifecycle use*
4. *Structural scanning*
5. *Site investigation for engineering and security planning*
6. *BIM verification*
7. *Large exterior sites*
8. *ADA compliance*
9. *Construction monitoring*
10. *Deformation monitoring (time lapsed)*
11. *Forensics*
12. *Historic documentation (including HABS/HAER[Historic American Buildings Survey/Historic American Engineering Record] documentation requirements)*

Comments:
Additional response received after workshop: No

Question 1C: Are these use-cases clearly stipulated in the SOW? **4 – Yes, 4 - No**

Comments: *Use-cases are not relevant for SOW. SOW document needs to be greatly improved.*

Question 1D: Is the project objective clearly stipulated (i.e., what GSA needs)

3 – Yes, 1 – No, 4 – Could be better

Comments:

1. *Minimum specified for scan resolution in addition to minimum artifact resolution*
2. *Clarification on deliverable format needed*
3. *The objective clearly states GSA's overall need, but should also include specific objective of the project (e.g., building documentation to support pending improvement project or develop BIM to support facility management and energy conservation initiatives). This would be helpful for the consultant as well as the local GSA project manager.*
4. *Additional response received after workshop: Could be better*

Question 2: Should GSA require a scan plan? **6 – Yes, 2 - No**

Comments:

1. *Needed to inform all primary stakeholders. Delays caused by questions/security checks increase field crew costs.*
2. *Scan plan appropriate to determine vendor's technical approach, understanding of scope of work. Scan location map in advance and calibration certifications not necessary at time of proposal. Ask enough to determine the expertise of bidder.*
3. *Scan plan helps define for all parties what is being covered and what areas are not.*
4. *Scan plans are important. Used to coordinate with security and building occupants. Can be used to better determine the cost of a project.*
5. *Not at the onset, but possibly at the end to get an idea of scanner occupations, also maybe use TrueView as part of deliverable.*

6. *Good idea. Often prepare a preliminary scan plan when quoting a job. A more detailed scan plan developed once arriving on site.*
7. *Additional response received after workshop: No. Not always needed. Often prepare a preliminary scan plan when quoting a job. More often scan plan cannot be followed because of access issues.*

Question 3: Is specifying a tolerance for a 3D imaging project the best way? **2 – Yes, 5 – No**

Comments:

1. *Meeting specified tolerance may not be possible due to software constraints. Better solution – specify a measure of fidelity, i.e, is the tolerance met given the limited technologies employed.*
2. *Accuracy, tolerances should be discussed and agreed upon – not scan resolution.*
3. *Area that needs to be discussed*
4. *Largest flaw in SOW. Approach technically incorrect.*
5. *Many ways to specify a project and this is more relevant to the application than an overall tolerance.*
6. *Additional response received after workshop: No. Many ways to specify a project and this is more relevant to the application than an overall tolerance.*

Question 4: What is the proportion (%) of costs for administrative (e.g., security clearance for personnel, permits, etc), scanning, post-processing data, and modeling for: Exterior, Interior – Offices, Interior – mechanical/utility, Historic, Sculpture?

This question caused some confusion. It appeared that some of the contractors took the question to mean "what was the administrative cost for each of the exterior, interior – offices, etc." and not "what was the cost (%) for administrative, scanning, post-processing data, and modeling for each of the use-cases (exterior, interior – offices, etc.).

The intent of this question was to help give GSA a better understanding of what the approximate costs are when determining their needs. For example, if the cost for modeling was 40 % of the project, the project manager will need to decide if there is a critical need for the model or not.

	Administrative (%)					Scanning (%)		Post-processing (%)		Modeling (%)	
Exterior	10	5	5	5	5 – 10 for non-gov't, 15 - 20 for gov't	22.5	45	22.5	25	45	25
Interior-offices	10	5	10	10		16.25	30	48.75	30	25	35
Interior - mechanical/utility	10	5	10	10		16.25	25	48.75	30	25	40
Historic	10	5	10	5		16.25	30	48.75	30	25	35
Sculpture	10	5	5	5		16.25	45	48.75	25	25	25
Structural	10					16.25		48.75		25	

Notes:
1. Each column represents input from a contractor. Only 5 contractors answered this question. Two contractors answered the question correctly. Two contractors only supplied response for "Administrative". One contractor only supplied a generic response for Administrative.
2. One contractor added the "Structural" category.

Comments:

- Additional response received after workshop: All types of projects:
  Admin = 5 %   Scanning=30 %   Post Proc = 20 %   Modeling = 45 %

Question 5: How close are the actual costs to the original bids? The difference between the original and actual bids is: < 10 %, (10 to 20) %, (26 to 40) %, > 40 %.

The intent of this question was not to determine the contractor's profit margin but to determine if cost overruns were problems. If so, were the cost overruns caused by misunderstanding of the SOW or a poorly written SOW and how this could be remedied.

< 10 %	10 % to 20 %	26 % to 40 %	> 40%
2	2.5*	1.5*	1
*The fraction was because one contractor had a split response.			

Comments:

1. Historically, over runs experienced can be associated with the actual modeling efforts. Many times, our clients (both government and non-government) are new BIM users, and really do not understand how the model is developed and can be used. Once the

5

Question 6: Are there instruments that are better suited for certain projects? **8 – Yes, 0 - No**

Comments:

1. *Depends on the application and accuracy levels needed.*
2. *Longer range instruments for exterior work. Interior work requires greater detail. Instances where the scanner operator has few minutes in a certain area (hospital above ceiling, nuclear) which requires a certain type scanner.*
3. *Time of flight for exteriors and large spaces, phase scanners for interiors, MEP and architectural detail - never rule out conventional surveying as an option.*
   - *This is the most accurate response with the most common tools available. However, newer/developing product will offer the range of a "time-of-flight" with the speed and detail of a "phased-based" scanners (e.g., Z+F 5010i).*
4. *Additional response received after workshop: Yes*

Question 7: How often (% of the total number of projects you have worked on) is it discovered that during the modeling phase, more scans are needed or scans had to be repeated?

> 80 %	> 50 %	< 20 %	0%
0	0	7	1

Comments:

Question 8: How much of a problem is interoperability between software packages?

0 – Large, 2 – Medium, 5 – Small, 1 – No Problem

Comments:

6

1. *Limitations of the current available software seem to be the biggest contributor to increased cost of modeling scan data.*

    *   *Very accurate statement. The fact that BIM by nature is an orthogonal system limits the ability to accurately depict the reality of the point cloud.*

2. *Lack of interoperability can be worked around.   Main issue is lack of backward compatibility of some packages.*
3. *BIM software has problems with raw point clouds.*
4. *Additional response received after workshop: Medium. Agrees with Comment 1.*

Question 9: How do you suggest that GSA check the following deliverables?

Point Cloud:

- Registration, GPS, traverse, and control reports
- Need to define standard registration error numbers
- Visually
- Overall dimensional and control checks
- Manual measurements (utilize fixed features or artifact of a known dimension)
- Cross sections (horizontal and vertical) through point cloud – check for gaps
- Check elevations at control points

2D Plans:

- Overlay 2D plans on point cloud
- Manual measurements/spot checks
    - Point-to-point/dimensions
    - Area

3D Models:

- Check against point cloud
- Manual measurements
    - Point-to-point/dimensions
    - Area
- Visually

Question 10: Excluding costs, what factors should GSA consider when evaluating bids?

The factors below were not ranked in any way; although Experience, Past Performance and Project Approach were factors that were mentioned by several contractors.

1. Spread in the bids submitted. Large variation = not everyone bidding on same scope of work. Bidders have sufficient time to bid the job? Opportunity for site visit and ask questions?
   - *Site visit is ideal, but not always practical due to end-of-year funding. If not an option, GSA should provide as much info as possible and conduct pre-bid teleconference.*

2. Experience and technical qualifications – field and office staff
3. Past performance: with similar projects and in delivering on time
4. Project approach
5. Scanning equipment used
6. Data security
7. Time frame for completion/schedule
8. Are there any assumptions made or qualifications for work which might impact deliverables?
9. Travel and mobilization
10. Security and location access
11. Modeling requirements
12. Firm: Size and background (surveying, architectural, engineering, CAD)
13. Is work done in-house or out-sourced overseas?
    - *Work should never be sent overseas/Canada (taxpayer's money)*
14. *Experience, past performance, project approach and equipment used were key to the IDIQ award, and therefore should not be a major factor in task order award.*

Question 11: Rank the list of topics for discussion during the workshop (with 1 being the most important).

Response:

1 = Scope and specification
2 = QA/QC (Quality Assurance/Quality Control) best practices
3 = Estimate
4 = Deliverables
5 = Rules of thumb

6 = 3D imaging lessons learned
7 = Metrics
8 = Post-processing workflow
9 = Future uses
10 = Other: Lack of use of IDIQ contractors, Schedule/Access, Lack of work orders

3 Likes, Dislikes, and Workshop Expectations

At the beginning of the workshop, the participants were asked to write down one like and one dislike of GSA's 3D imaging program, one suggestion for improving the SOW, and their expectations for the workshop. The likes, dislikes, and suggestions for improvements were to have been discussed at the end of the workshop but the discussion was omitted due to a lack of time. The un-edited responses are given in the following four sections.

3.1 Likes

1. National focus on use of digital technologies for building documentation.

2. GSA program is very broad and open in possible use of laser scanning. The concept of using point cloud data to support activities other than BIM development (i.e., facility management/cmms [computerized maintenance management systems]) has been well received.

3. Willingness to change.

4. That GSA is trying to utilize advanced technologies like laser scanning.

5. Program expands the consideration of using laser scanning in buildings and builds support for the technology.

6. GSA's push towards adopting performance metrics for evaluating project outcomes. (CMU [Carnegie Mellon University])

7. GSA's leadership in the industry is moving the technology forward.

8. Pre-qualification of teams/persons.

9. Pre-qualification of 3D laser scanning vendors.

10. Involvement helping GSA further its scanning/BIM initiatives; serving the USA; involvement with other players shaping our industry.

11. Forward looking stance of the contract and hope it will push innovation forward.

12. We like the way GSA attempts to level the playing field for all vendors.

13. The proactive role GSA is taking with laser scanning.

14. Willingness of Region POCs [Point of contact] to learn about the technology and look for ways to leverage it.

15. I like the fact that GSA is bringing diverse people together from academia, government, and industry, and that they are providing the impetus for moving this field forward at a rate that is faster than would be done otherwise.

16. Progressive utilization of the technology. Paving the way on how the laser scanners can and should be used for commercial buildings.

3.2 Dislikes

1. Work packages are competitive fixed price contracts which was not the intention of the IDIQ – too much risk.
   - *Disagree. Most IDIQ contracts are for fixed-fee (lump sum) work orders.*

2. Communication seems to be very limited. Many regional staff members have not heard of/seen laser scanning, which results in a lot of confusion.

3. Slow movement on task orders.

4. That GSA is not compiling case studies around the thousands of 3D laser scanning projects that are happening by commercial firms around the country every year.

5. The lack of work orders casts doubt on the true value of laser scanning and whether it is applicable and affordable.

6. Difficulty in bidding – lack of site visits pre bid, limited pre bid documents available.
   - *Key issue*

7. Separation from A/E (Architect/Engineer) contracts. A/E contractors hire without regard for IDIQ holders.

8. Lack of regular use of contracts.

9. Lack of use of IDIQ contracts.

10. Turnaround time to quote jobs.

11. I do not feel like most of the regions are aware of the value and more needs to be done to sell them on the contracts.

12. What funds are available.

13. We would really appreciate a greater insight into how we can get more involved with project managers to insure the SOW does not limit the IDIQ contractors since the sheer size can be daunting.

14. Due to the lack of projects released in Region 10, we do not have enough experience with GSA projects to provide valuable input.

15. Small number of task orders on the contracts.

16. I think that we could benefit from a more aggressive transition of the research that we are doing to general practices.

17. The rationale behind some metrics of accuracy is not very clear.

3.3 Suggestions for Improving Scope of Work Document

1. Improvement in coordination and understanding between GSA and IDIQ contractors.

2. More direct involvement with contract PMs.

3. The GSA needs to get out and watch 3D laser scanning projects in action from start to finish.
   - *This is an excellent piece of advice. I highly recommend this.*
   - *Best comment in this proceedings*

4. Require A/E contract holders to use IDIQ (3D laser scanning) contract holders.

5. Clearer RFPs and longer response times.
   - *This would be very helpful. It is clear from the SOW's that there seems to be a great deal of time to put an SOW together yet the response time has barely been sufficient to even understand the SOW.*

6. The primary improvement – the program needs a consistent way to develop project specific requirements. Every project is different, and the process is more important than a SOW or specification.

7. Develop a shorter form for PMs to help guide them in using the IDIQ contractors.
   - *Great idea*

8. Release at least one project per Region so each Region can have a real project with which to work and test.

9. Suggest use of data standards be embedded in SOWs as guidance.

10. I would dedicate more resources to this project so that we can increase the pace of advancement of this area. I would like to see the US as the world leader in this area.

11. Having more feature driven requirements on the accuracy of the data collected and model/information generated.

12. *I read through the proceedings and I would totally agree with the comment that the SOW needs to be greatly improved. The scope of work needs to outline specifically what the final intention of the model will be to determine the best practices to reach that goal. If the final model is for general space planning, that takes a different approach then a model that will be used for clash detection and routing of MEP systems.*

    As for level of detail, specifying "artifact size" is not an appropriate way to determine density of point cloud for the entire project. Specific items needed by the end user should be listed in the SOW so the laser scan team can then determine the density based on the need.

3.4 *Workshop Expectations*

1. To gain a better understanding of where GSA would like to go in respect to users (short and long term). Also to understand QA/QC practices used by GSA.

2. How can I be more proactive in defining SOW and movement on task orders.

3. Awareness of our fantastic 3D scanning projects we do every day around the world.

4. Plans, suggestions, and options for generating more work orders for these types of projects.

5. I would like to learn towards which direction GSA is heading in terms of shaping future of scanning program; more standards involvement, more research and in what area, etc.

6. Clear understanding of where contract is going.

7. Better understanding of GSA's intentions for IDIQ contracts.

8. A better understanding of GSA's SOW and goals.

9. Understanding of current IDIQ – went through interview process, how it seems original, IDIQ process is open and not following requirements and contract indicated by GSA.

10. Status of on-going (DRAFT) standards efforts to support SOW – i,e: ROBCAD, etc; Nature of feedback from GSA Regional Project Managers to Central Office regarding value of scanning/modeling; Opinions from contract holders always interesting; Learn where scanning/modeling is showing the most value to GSA.

11. I am hoping to learn more about the needs of the contractors so that we can better identify the areas that need more research.

12. Outside of technical requirements for the performance specifications, I'm hoping most to get clarity on accepted use-cases & their respective workflows from data capture – registration – to ultimately BIM integration for the laser scan use-cases. Also, I'm interested in the discussion regarding the various technologies that can serve these workflows.

General comments on Section 3:

- *After speaking with a number of GSA project managers and witnessing discussions in the GSA BIM conferences I've attended there seems to be a common theme developing in what I am hearing regarding the view of the IDIQ contracts. Project managers have been stating that they don't intend on using the IDIQ contracts. Reasons given include:*

  - *Not having the time to create a SOW or learn how to create an SOW.*
  - *Not wanting the responsibility of managing another contract*
  - *Not wanting to be responsible for the laser scanning contractor should something go wrong. If they just make the A&E responsible for any required as-built documentation they don't deal with any of the above concerns.*
  - *The IDIQ's are just a contracting vehicle with no budget. They still need to find and allocate funds.*

  Other observations include there seems to be a certain percentage of GSA personnel that are not interested in learning about the scanning/BIM technologies. This could be why there doesn't seem to be many task orders being issued.

  Not sure how to resolve this. Maybe if there were some way to allocate some funding to go along with the contract, maybe more PM's would start to use it.

  I am confused as to why such a stringent vetting process was used to award the IDIQ contracts, but there doesn't seem to be any vetting process for scanning/BIM contractors who are being utilized through other contracting vehicles, or who are hired by an A&E firm working for the GSA.

- *I very much liked the request early in the day for the "Likes/Dislikes/Expectations" responses. I liked that they solicited feedback in an anonymous way - even though we didn't have a lot of time for the responses.*

4　Brainstorming Session on the Scope of Work Document

In the brainstorming session, the participants were given 20 minutes in which to individually write down ideas about the following SOW sections:

1. Project Information (e.g., Background, Objective statement, Description)
2. Proposal requirements (e.g., Scan plan, cost breakdown)
3. Performance metrics (e.g. tolerance)
4. Deliverables (e.g., point clouds, 2D plans, 3D models, format for current and future access)
5. Acceptance criteria (e.g., X % field measurements out-of-tolerance)
6. Others

Each idea was written down on a post-it note and stuck under one of the sections listed above on a whiteboard in the meeting room. After the brainstorming, each idea was discussed to ensure that everyone understood what was intended and clarifications were sought (if needed). Within each SOW section, the ideas were further grouped into sub-categories if applicable (see Figure 1).

The whiteboard is "reproduced" in Sections 4.1 to 4.6. In Sections 4.1 to 4.6, the text in black is the information as written down by the contractors on the post-it notes, and the text in blue is added to help clarify the idea and is based on the general discussions about the idea. The numbers within the "{ }" is the number of post-it notes with the same or similar idea. This gives an indication of importance and/or agreement of the participants with the idea.

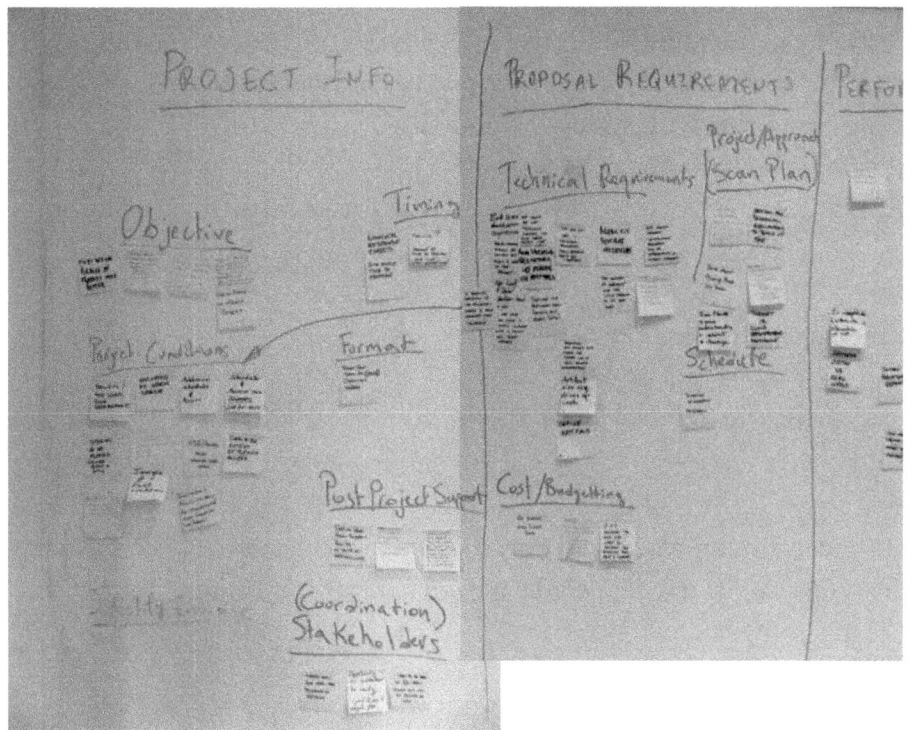

Figure 1. Photo of a portion of the white board with brainstorming ideas posted for Project Information and Proposal Requirements.

4.1 Project Information

The ideas under Project Information section were grouped into seven sub-categories: Pre-bid process, Project Conditions, Objective, Timing, Format, Stakeholder Coordination, and Post Project support.

a. Pre-bid Process
   1) Site Visit {**4**}
       a) Determine accessibility
       b) Get more information to prepare bid
   2) Need better process to discuss project requirements
   3) Phone conference to supplement SOW to better understand project needs
b. Project Condition – needed information for bidders
   1) Occupied space vs. what is free space
   2) Security and work hour restrictions
       a) Areas that require escorts

17

b) Areas that cannot be imaged – not because of technical limitations but because of tenant's restrictions

3) Current electronic file for the facility
4) Specify the number of floors – include basement and attic
5) Images, photos, drawings are helpful **{2}**
6) Describe extent of plenum access
7) HSE (Health Safety Environment)/**Access issues impacting field work** – things that you have to deal with such as asbestos, scaffolding not allowed, and confined spaces.
8) Schedule and access **{2}**
   a) When is space available
   b) Having to skip an area/office and having to come back later reduces efficiency of work flow

c. Objective
1) Good description but should only include those that are relevant to the project … the scope we have seen include all, and leads to confusion
2) Clearly define area of interest. Should include graphic such as site map, elevations, and floor plans
3) Must define purpose of project much better
4) Need to understand project objectives including project turn-over to facility management for interoperability
5) Define need to match budget
   a) PMs ask for everything and wanting to pay for a little. Cannot ask for 100 % coverage if only have budget for 20 % - this needs to be up front.
   b) GSA PMs need to estimate budget for a project and needs some feedback (rules of thumb) from the contractors. Suggestion to use past projects as a guideline.

d. Timing - **Need to allow for longer response/turnaround time for bid {3}** – Not possible at times due to the government procurement process for end of fiscal year. GSA may find that they have some funds one month before the end of the fiscal year and it has to be allocated before that or the funds are lost.

   • *In circumstances where limited time can be given, suggest putting out a notification that an SOW is being developed and to be prepared for its issuance and estimated turnaround time. This will at least give the IDIQ contractors some sort of heads up that something is coming and to clear their schedules to better be able to respond in the limited timeframe. It isn't uncommon to be traveling or otherwise unavailable when an SOW comes in. Knowing in advance that it is coming can help us be prepared to work with the shorter timeframe.*

e. **Format – short form for IDIQ contractors.** Contractors already pre-qualified through IDIQ process do not need to go through a re-qualifying process again. GSA agrees with this. Current SOW was prepared for generic 3D imaging contract.

f. Stakeholder Coordination

1) **Varies with end user. Need discussions with A/E teams.** What is the deliverable format? Need coordination with all design team – e.g., contractor models all the joints as required, but A/E team does not need the information and throws it away.

2) **Opportunity for contractor to verify conditions and adjust plan.** For unexpected situations (e.g., wall not parallel) contractors should have the opportunity to comment/document that the conditions were not as expected and to change the scan plan or have change orders.

   - *This is a reasonable request. Especially if a pre-bid site walk was not possible. Seems fair that upon award and at the time of the initial site walk any adjustments could be made upon visual discovery of actual conditions unknown at the time the bids were prepared. It is important to note that the people putting the SOW together for bid are already pretty familiar with the SOW. The contractors have no knowledge of the facility, typically, when that SOW hits their desk. Graphical information in the form of record drawings and photographs are extremely helpful when putting together a bid. Providing this type of information is more likely to result in better bid numbers because the risk factors are diminished.*

3) **Need to be part of A/E team. Project cannot stop at the delivery of data.** GSA needs to inform A/E firms that they (GSA) have IDIQ contractors that are available for their use, if needed. Response - GSA cannot specify which contractor an A/E firm hires. Other agencies are asking about GSA's BIM and laser scanning IDIQ contractors and GSA is looking into how other agencies can use their IDIQ contractors without their having to go through another screening process.

g. Post Project Support

1) **Define post scan support (pg. 34 of GSA's SOW for 3D-4D-BIM Services document) or make as additional work** – post scan support is very open ended.

   - *This is a tremendous risk for all bidders and, quite frankly, we would never agree to this in our private sector work without some sort of definition as to what the support entailed or at least a definition of a set number of hours that are included.*

2) Post-project coordination budgeting and scoping of what is truly expected

3) Section 9.4 Post Scan Support is very open-ended. Not possible to provide "hard" costs. Consider adding a set number of hours in addition to 6 month period.

   - *Good comment. I hope this is given serious consideration.*

4.2 Proposal requirements

The ideas under this section were grouped into the following sub-categories: Technical Requirements, Schedule of Completion, Cost/Budgeting, Technical Approach/Scan Plan.

a. Technical requirements
   1) End users should determine requirements
      a) Table 1 (in the SOW template) with three options is not a good way to develop what those requirements are. If the requirement is spatial program validation, then the tolerances in the table are higher than what it needs to be. For construction fabrication, the tolerances are too high.
      b) The end use should be considered. What do you need? This should drive all the requirements. There has to be an immediate need. One suggestion was to have Table 1 list use-cases and associated with each use-case is a tolerance or resolution. Or a grouping of tolerance and resolution and use-cases tied to each grouping.
      c) GSA needs to understand what they are asking for and the associated tolerances needed. The contractors cannot bid on future uses. Verbal discussion about the project needs – pre-bid conference call with GSA PM.
   2) A use-case work scope is limiting compared with a project goal-based approach. Need to differentiate between use-case and end use. Picking from a list of six use-cases is very limiting.
   3) A detailed description of the architect's needs is more important than tolerances. Many of the topics in the Project Information and Proposal Requirements sections could go in either section – the line between these two sections is very blurry.
   4) The use of "etc." in describing requested deliverables should be removed.
   5) Break out features into categories – e.g., requirements different for MEP and for walls. Look at different building scopes and see what is required.
   6) RFP should address "compromises" requested such as orthogonality or uniform components. Scope of work does not address this issue. Issue with Revit - if you step out of orthogonality in Revit with MEP systems - it is 30 % to 50 % more hours.
   7) The concept of artifact size has little meaning to an end user.
   8) Define difference between scanning and modeling tolerances.
   9) Define line between scan density and model detail.
      a) This goes back to requiring 100 % coverage and asking for 1 in. x 1 in. artifact - the ability to go back on a project (future uses). The real value of what we do is the raw data – it is not the deliverable for a particular project. It is what we can do with it 10 years in the future which is not the same in all cases. The raw data is based on what you want to do in the future – what is the line between what is

needed now and what is needed in the future. Maybe having a minimum of what is needed in the future would help so that the PM does not have to guess what is required.

    b) Scan data is the better value – the other stuff is added cost. By having the "added" cost, we (contractors) don't get the contract – the cost is too high.

         * *Disagree with this comment, but understand the intent. Perhaps GSA should utilize task options (Scanning, Modeling, Support) and consider issuing options to one or multiple contractors based on costs.*

  10) Avoid specifying technology methods. Let experts use best tools.

       * *Disagree. GSA need to identify what their expectations are, and what the deliverables need to be. As consultants, we have the responsibility to identify when something better can be employed and bring those to the attention of the GSA, but GSA goals must be the starting and end points.*

  11) RFP should be less technology centric and more needs centric. Let bidders explain their approach.

  12) Clear level of detail definitions based on use.

  13) Requirements should be driven by goals and needs of the project.

  14) Define artifacts.

  15) Artifact size requirement drives up cost.

  16) Specifying raw artifact size misses the target use of most building documentation

b. Schedule of completion for field/office – do you want it done in 1 month or 5 days? What is the project schedule?

  * *This is important since we can staff our projects differently to meet various schedule requirements. If we propose a timeframe when one is not provided in the SOW, please consider that every bidder may come back with differing timeframes within their proposals due to how they plan on staffing the job. Timeframes for delivery can often be reworked to meet specific project requirements.*

c. Cost/Budgeting

  1) Firm fixed price generates too much risk without airtight SOW especially when Revit is involved – e.g. "etc.", post project 6 months support requirement

  2) Cost breakdown should be as detailed as possible: mobilization/demobilization, coordination w/ personnel, field work, registration, point cloud, 2D, 3D modeling, etc.

  3) Fee should stay lump sum

d. Project approach/Scan plan - if a scan plan is required, need to define at a minimum what goes into the scan plan

  1) Scan plan to prove understanding. Scan plan is subject to change.

  2) Target vs. cloud registration guidance - the type of targets, registration affects the outcome.

3) Scan plan should include scan worlds, safety, traffic control, instrument stabilization.
4) Build project planning phase into Scope.
   a) SOW jumps right into scanning. There is no time built into SOW for talking to PM to get a better feel of what is needed outside of the actual requirements.
   b) This would be post-bid. Are you willing to pay for that part of the planning?
5) Scan plan of methods should be reviewed as a "selection" for award of contract
6) Section for technical assumptions to back up fee

4.3 Performance metrics

a. Tolerances are acceptable. Coverage causes issues.
   1) By specifying 1 in. x 1 in. or 2 in. x 2 in. (assuming 100 % coverage) requires dramatically more scan locations, and you are getting unnecessary objects. Future needs cause problems with what the current needs are. The problem is not the 1 in. x 1 in. but the 100 % coverage – e.g., flat wall 100 % coverage is not necessary – flexibility needs to be incorporated into the SOW for 100 % coverage.
   2) Without laser scanning, similar work has been performed in the past. What is different now? Laser scanning is not different from traditional methods – it just does it faster and more accurately – writing the SOW should not be any different than it was the past. As long as the minimum requirements (immediate need) and add on requirements (future needs) are not based on technical requirements but based on how it will be used.
b. Need to address orthogonal vs real world.
   • This is critical, especially when dimensional tolerance requirements are made part of the contract. It may not be possible to guaranteed specified tolerances due to the current software technologies.
c. Products should adhere to 95 % confidence level so 5 % out-of-tolerance seems reasonable.
d. Performance metrics must evolve as technology improves.
e. Control and registration reports.
f. Is simplification (orthogonalized etc.) allowable or not?
g. Absolute vs. relative accuracy – what coordinate frame of reference is used when determining out-of-tolerance.
h. Specify objects (lights, walls, molding, etc) required and level of detail for point cloud/model.
i. What about tolerances for occluded region % of point density?

4.4 Deliverables

a. Data format/standards
   1) ROBCAD effort shows great promise to help project teams speak the same language and bridge gap between as built data collection & model development and design. Looking at standard way to specify how equipment is named based on national CAD standard, omni class standard
   2) Earth centered, earth fixed coordinate system. For a 3D imaging project for an entire campus, it is a challenge to define a global coordinate system. What coordinate system to use? Google and Microsoft are moving towards earth centered, earth fixed – it has the advantage of portability for web services.
   3) ASCII as a deliverable format is inefficient.
   4) BIM interoperability and facility management.
   5) ASTM E57 file format as "encouraged" deliverable (similar to GSA's IFC initiative)
   6) ASTM E57 data format for scan data
b. Define different requirements by scope (Bldg. scope) MEP, steel, etc. Deliverable requirements should be defined by scope. There are requirements for electrical, etc.
c. Completely define the expected deliverables (e.g., is it truly necessary to model in greater detail than the design requirements?) Clearly define the deliverable
d. There should be a clear distinction between geometric and parametric deliverables and why used. Make sure you understand what and why you are asking for something. Is this issue a part of the technical requirements or deliverables?

4.5 Acceptance Criteria

a. Error/Uncertainty checks
   1) Truth vs. scan/2D/3D/BIM {2}
   2) X number of verification measurements per square foot or number of scans. The means of verification is very important. There was ambivalence as to the question of "How do you verify that?" There was a suggestion to use a calibrated object.
   3) Cross section delivery for visual verification – visual check of model vs. cloud.
   4) Include standard device/artifact in the environment – e.g., scan artifact every X number of scans.
   5) Deviation report of the scanned and 3D and existing – scan vs. 3D and scan vs. existing.

b. Omission tracking - Analytic model to data comparisons (e.g., current work at CMU[1]) will be the future of QA. This is the future of QC. Compare the point cloud to 3D model. Omission is more important than accuracy – if my job is to model MEP systems, and if I missed a whole bank of ventilation systems, then this is more important than if the system is out of place by 1/8 in.

c. Allow 2D scan data instead of requiring PBS CAD standard. Similar to requiring a model, you can cut pretty simple plans out of point clouds without having to build something more complex. Is there a way to work with the point data? The PBS CAD standard (for 2D) is too stringent. It may be more time intensive than value added. There is no standard/accepted way to use the point cloud to get 2D plans (project all the points for a wall to create a line, etc.). Need more work on defining what is acceptable for 2D deliverables out of laser scanning. Currently, there is only one way (PBS CAD standard) and it is too stringent.

d. Require QA reports that tie to scan plan and model plan. Define contents of this report.

e. It is possible to build check lists to help acceptance. There are standard check lists based on technical requirements. There currently is no method for acceptance. Could be based on best practices – we used this procedure, and followed that, etc.

f. Should be performance based. GSA agrees but how?

g. Spot measurements for performance checking lead to conflict and are not a good performance metric.

- Checking a couple of spots here and there – absolutely not – leads to litigation. The measurements from a tape measure or Disto[2] are not as good as those from a laser scanner.

- Require some kind of metric to be placed in the environment to verify the accuracy of the 3D imaging system.

- If the 3D imaging system is calibrated, it is more accurate than a tape measure; have seen this to be true in case studies. It was pointed out that this misses the point of the use-cases – if orthogonal models are needed, it destroys the linkage between the data and the truth. Current software does not have the capability of representing the real world.

- Like a total station, a 3D imaging system is calibrated and the operator is competent, you don't check the distance obtained with a tape measurement – you accept it because you trust the calibration. There was a cautionary note that surveying uses

---

[1] Anil, E. B., Tang, P., Akinci, B., and Huber, D., Assessment of Quality of As-is Building Information Models Generated from Point Clouds Using Deviation Analysis. May be downloaded from: http://dl.dropbox.com/u/13678634/2011-anil-spie-qa-final.pdf

[2] A Disto is a handheld, laser-based distance measuring device.

prisms and obtains few points while in laser scanning there are many points and the quality of the points depends on how you scan.

- If have calibration reports, 99 % of the errors are due to poor registration – you are not talking about ¼ in. but 4 ft. Point is spot-checking where a novice user wants to check you in one or two places, it is a lot more complicated – they can check using spot checks but they cannot contract to that.

4.6 Others

a. Workshop objective (Educating GSA PMs)
1) Break out and define in the bid submittals a few areas that always come back as a concern
2) Section 7.4 is very weak. Need to clearly define and communicate the requirement and format/content of the evaluation report
3) Contract holder directly involved in development of SOW.
   - Need to involve contractors because project managers do not understand what they want.
   - Allow IDIQ contractors to assist PMs through workshops, conference calls, direct contact to educate them on what to ask for. GSA agrees.
4) Understand what you are purchasing – more education. GSA PMs do not understand what they are getting or asking for.
5) Method to have multiple vendors assist in SOW w/ project managers to assist them (multiple vendors ensures it is an even playing field)
6) Involve contractors in budgeting
7) Tolerance is not an exact unit of measure – how else to specify performance metric?
   - *Maybe specify a process to be used with some form of verification that the process was performed properly. The focus would then be on adopting a process that would ensure a high level of confidence in the outcome, yet takes into account the limitations of today's current technologies and not requiring the contractors to have to agree to something that may not be possible to guarantee through no fault of their own.*

b. Edit/improve language in SOW template
1) Widen the array of acceptable explicit methodologies included. Primarily mobile vs. static scanning. GSA is not restricting technologies. If anyone sees this in the SOW template, please let GSA know. The restriction seen is in the use of the term "laser scanning", that is, calling for the use of laser scanning precludes other 3D imaging technologies. Suggest use "3D imaging" (as suggested by ASTM E57) instead of "laser scanning".

2) BIM requirements should reference BIM (5.0) section. The BIM section 5.0 is more detailed.

3) Improve overall organization [flow] of document. Sections are broken up where similar things are found in different sections.

4) Use better samples as examples. Better SOW.

5) A template for documentation types is useful and common. E.g., a go-by for exterior elevation content and detail. Graphics are easy to understand.

c. 2D plans … OK to ask for, but should be extracted from 3D model.

1) A suggestion was to go from point cloud to 3D then to 2D. There was disagreement about this. A contractor stated that 80 % of what they have done over the last decade is 2D deliverables. Software limitations – documentation in Revit – not happening. For ornamental buildings, 2D drawings have to be generated - cannot go from 3D to 2D.

   • *This is true, but it does not meet GSA's overall objective.*

2) There is appropriate need for 2D. 2D should not be neglected or considered low-level technology. There are areas where 2D cannot be done without 3D imaging.

5 Changes to Scope of Work Document

5.1 *Summary of Discussions*

After the clarifications and classifications of the ideas, the workshop participants decided to focus the remaining time on two SOW sections: Project Information and Proposal Requirements. It was felt that if these two areas were correctly specified, the other sections of the SOW would also be prepared correctly. However, the discussions touched on most of the other SOW sections.

Some of the main points from the discussions are summarized below:

1. *Site visit.* The contractors felt that site visits were essential to help them prepare their bids. The site visit helps them to:

    a. Determine lines-of-sight and technical approach

    b. Determine areas of congestion and occlusions

2. *Pre-bid information.* Having the following information would be very helpful to the contractors:

    a. Electronic plans/drawings.

    b. Images, photos, drawings of site and areas of interest/importance/high level of detail required.

    c. Other projects that may be occurring in parallel that may require coordination with the 3D imaging work.

3. *Pre-bid conference*: The contractors felt that there was a need for a pre-bid conference with the GSA PM and A/E firm (if known or available) so that they could better understand the project requirements.

    The pre-bid conference would allow for discussion of project requirements and conditions such as:

    a. Some access issues that have arisen in past projects include:

    1) Security escorts required – at all times, after hours only, only for certain areas?

    2) Can only work at certain hours?

3) Work had to be conducted out of sight of the public.

4) Can only scan above certain height from the ground and below a certain height from the ceiling?

5) If an area is not accessible when the contractor is there and the contractor has to come back (either another day or at other time during the same day) makes for an inefficient work flow and causes delays.

   - *This is a critical constraint when project teams travel for the field work. Typically during negotiations, contractors are required to limit the number of days/nights in the field to control per diem costs. Any delay can impact those cost factors as well as airline change fees, rental vehicles, etc.*

b. What needs to be imaged? How will the data be used?

c. A/E firms need to be involved so that they can specify what their needs are and also if the data needs to be in a certain formats for their software/use.

d. Is there asbestos involved?

e. Is imaging above the ceiling required?

4. *Time to prepare bids*. The contractors stated that at times they were given very little time to prepare their bids, and they would like to have more time.

5. *IDIQ contracts*.

a. The IDIQ process for contracting work did not seem to be clear to the contractors (and maybe also to GSA PMs). The contractors felt like they had to go through another qualifying process to get the contract/task when they already went through the process to be qualified as an IDIQ contractor. The contractors expressed the need for a shorter form/process for IDIQ contractors due to the pre-qualification. GSA agreed. The SOW template, as currently written, was developed for preparing SOWs for generic contracts (i.e., non IDIQ contracts) and needs to be changed.

b. Are the contracts "firm, fixed price" or "time and material"? Whether the contract is firm, fixed price or time and material is up to the individual GSA PM. The contractors felt that firm, fixed price contracts introduce too much risk for the contractor. Also, for these types of contracts, the SOW needs to be cleaned up and the scope of work has to be very specific. The SOW should not have "etc." in it and the post-project 6 month clause needs to be re-worded (e.g., consider adding a set number of hours).

   c. There were several comments regarding the lack of work for IDIQ contractors.

6. *Performance metrics.*

   a. Table 1 in the SOW template: The use-cases are too broad and need to be narrowed down. Suggestion was to specify tolerances for each use-case, but this is probably not feasible.

   b. Use a table similar to that shown below where the cells will be filled with the required deliverables based on the stakeholder and tolerances would be associated with the deliverables.

Stakeholder/End User	Need/End Use					
	MEP Documentation	MEP Coordination	Space Mgmt (Space Documentation)	Historic Documentation	Urban Planning	Art Documentation (Exterior)
GSA Design and Construction						
GSA Facility Management						
GSA Portfolio						
AE						
Contractors						
Tenants						
Public/Preservation Community						

   c. Instead of specifying tolerance, specify components and exclusions. For example, pipes over 1 in., hangers, brackets, all pumps, etc. Specify a list for a geometric model and a separate list for BIM model. This would be in line with the performance-based approach that GSA is striving for in their contracts.

A word of caution that was brought up was that if left up to the contractor, the data density may be determined to be insufficient due to the contractor's mistake or inexperience. However, this could be mitigated by reviewing the contractor's past experience and performance and contractors making these types of mistakes will not be in business long.

     • *The bigger issue here is the ability to re-use the point cloud data for future needs. The point cloud may be detailed enough sufficiently document the existing condi-*

tions, but not for facility management uses such as CMMS integration or Space Utilization.

d. Some contractors indicated that coverage may be more important than specifying a tolerance. A contractor indicated that he imaged all projects at the same point density (regardless of project requirements) and therefore coverage would be more important.

   - *This may be the case for one contractor, but not for all. GSA needs to identify their standard for cloud density. If a contract exceeds that standard that would be acceptable, but a minimum needs to be identified to allow re-use of data.*

e. Some guidance for required tolerances was given: 1/8 in. for construction, 1/4 in. for design, 1/2 in. for others (e.g., planning, mapping).

7. *Acceptance criteria.*

a. As indicated in the pre-workshop responses and in some of the discussions during the workshop, manual measurement/spot checks were suggested as a method for checking deliverables. However, there was a feeling that the manual measurements obtained with a measuring tape or Disto are far less accurate than the measurements obtained from 3D imaging.

b. There was a suggestion that having the 3D imaging calibration reports, registration reports, and survey report of control points would be sufficient to ensure the accuracy of the point cloud. Have a check list with which to ensure that best practices were followed and the necessary reports are available.

   - *These reports are a true test of data accuracy and are a key check for Quality Control*

c. Include a calibrated artifact or an artifact with known dimensions in the scene. This artifact would be measured every X number of times.

   - *This is a very good practice, but it is important to note that the "artifact" should be positioned so it may be easily removed from final/processed point clouds.*

d. Simplification due to software restrictions: Dimensions could be out-of-tolerance due to simplifications imposed by the software. Some examples are being walls being made parallel or planar when they are not and walls being made orthogonal when they are not. This issue is known and needs to be addressed.

30

e. Use deviation analysis as proposed by CMU (see footnote 1). Statistically or visually compare the deviations of the point cloud and the model. Visually check for omissions.

f. How are the deliverables from A/E firms checked? Ask A/E firms for their input on how to check deliverables.

   • *Assume this is referring to a "peer review"... if so, GSA should be cautious and relationships between firms may compromise the effectiveness of the process.*

g. Licensing or certification of contractors: The contractors did not seem to want this.

8. *Deliverables.*

a. Point cloud: Scan data or raw data may be the better value. Sometimes the point cloud is all that is needed. The other deliverables (e.g., 3D models) can dramatically increase the cost and the project is not awarded.

b. Format: Having a standard data format is desirable to both contractors and GSA as it enables interoperability and is important when the data may be needed 5 to 20 years in the future. The standard for the ASTM E57 format should be available by Sept. 1, 2011. Software vendors such as Trimble, Leica, Faro, Pointools, and Optech have indicated that they will have the E57 format available in their next release of their software. It was suggested that GSA include in the SOW a requirement such as "E57 data format is preferred" as a first step toward making the E57 format a required deliverable.

9. Some contractors felt that doing smaller 3D imaging projects would make GSA PMs more comfortable with 3D imaging and it would make the PMs more knowledgeable about 3D imaging. Also, the current SOW is not written for smaller projects.

10. The contractors felt that the use of the phrase "laser scanning" restricted their choice of technologies. GSA stated that this was not the intention and the phrase "laser scanning" would be replaced by the phrase "3D imaging". The participants agreed that this change was appropriate.

5.2 Common Themes for Improvement

Based on the discussions during the workshop and from comments received after the workshop, some common themes (not ranked in terms of priority) for improving the SOW are:

1. Site visits are important.

2. Pre-bid conferences between PMs, contractors, and A/E firms are important.

3. Allow more time to respond to and prepare bids.

4. Instead of specifying tolerance, specify components and exclusions. For example, pipes over 1 in, hangers, brackets, all pumps, etc.

5. The requirement for PBS CAD format may be too stringent for 2D point cloud data. Another method or acceptance criteria needs to be developed.

6. Write the SOW based on current needs. It is very hard develop a SOW based on future needs. The need to anticipate future uses (i.e., require a higher density for areas that do not currently need the higher density) and the uncertainty associated with the future needs drives up the cost of the project, and this may lead to the project not being awarded.

7. Have GSA include in the SOW that the E57 format is preferred for 3D point data.

8. 2D deliverables should not be overlooked or deemed as low tech. A lot of the work that the contractors have done for other clients call for 2D deliverables.

6 Next Steps

The suggested next steps were:

1. Take categories on whiteboard and break them into webinar discussions or sessions.
2. Have IDIQ contractors take a cut at editing the SOW. Send any comments via email to GSA. Then meet in person or have a focused group meeting.
3. Have a contractor use a prior GSA job and re-write the SOW. What would he/she have liked to have known from Day 1 of the project?
4. If IDIQ contractors have better examples for SOW, send them to GSA.

Appendix A

List of Workshop Participants

	First Name	Last Name	Company
1	Peggy	Yee	GSA
2	Shane	Loyd	The RLS Group
3	Jody	Lounsbury	CHA
4	Kevin	Kianka	Stantec
5	Michael	Raphael	Direct Dimensions
6	John	Russo	Architectural Resource Consultants (ARC)
7	Eric	Hoffman	Quantapoint
8	Chris	Zmijewski	Stantec
9	Kelly	Cone	Beck Group
10	Brad	Adams	Woolpert
11	Engin	Anil	CMU
12	Matineh	Eylopoosh	CMU
13	Marek	Franaszek	NIST
14	Mitch	Schefcik	Cloudwise
15	Gary	Sheets	Coign Asset Metrics & Technologies
16	Calvin	Kam	Stanford/GSA
17	Kamel	Saidi	NIST
18	Gerry	Cheok	NIST

Call-in Participants:

	First Name	Last Name	Company
1	Ron	Aarts	Innovtec
2	Burcu	Akinci	Carnegie Mellon University
3	Daniel	Chudek	Innovtec
4	Joe	Davis	Custom Engineering
5	Troy	Day	Coign Asset Metrics & Technologies
6	Greg	Garner	PBS&J
7	Richard	Gee	GSA
8	Steve	Hagan	GSA
9	Daniel	Huber	Carnegie Mellon University
10	Eric	LaBrie	ESM Consulting Engineers
11	Bryan	Merritt	Erdman Anthony
12	Michael	Olsen	Oregon State University

13	Sam	Pfeifle	Spar Point Group
14	Gene	Roe	Lidar News
15	Terrence	Rollins	GSA
16	Stacy	Scopano	Trimble
17	Scott	Shin	GSA
18	Rick	Thomas	GSA
19	Rene	Van Kersbergen	Pharos
20	Ben	Williams	GSA

Appendix B

Workshop Agenda

8:45 - 9:00	Lobby Sign In
9:00 - 9:50	Welcome and Introduction

- Review of workshop objective
- Review of pre-workshop survey results
- Input from service providers on GSA laser scanning program

9:50 - 10:00	*Break*
10:00 - 11:30	Work session 1: Brainstorm about GSA Statement of Work

- Brainstorm suggestions for improving SOW
- Categorize brainstorm results
- Discuss suggestions

11:30 - 12:30	*Lunch*
12:30 - 2:20	Work session 2: Discussion of changes to current GSA SOW
2:20 - 2:30	*Break*
2:30 - 4:00	Work session 3: Discussion of additions to GSA SOW
4:00 - 4:45	Work session 4: Discussion of inputs from the introductory session
4:45 - 5:00	Suggested Next Steps
5:00	Adjourn

Appendix C

Pre-workshop Survey Questions

1A. GSA has identified several use-cases for laser scanning
   a. Exterior
   b. Interior – offices
   c. Interior – mechanical/utility
   d. Historic
   e. Sculpture

Are these the main ones?
   o Yes
   o No

1B. Are there any missing use-cases?
   o Yes (if yes, what is missing)
   o No

1C. Are these use-cases clearly stipulated in the Scope of Work?
   o Yes
   o No

1D. Is the project objective clearly stipulated (i.e., what GSA needs)?
   o Yes
   o Could be improved
   o No

2. Should GSA require a scan plan?
   o Yes
   o No

3. Is specifying a tolerance for a 3D imaging project the best way?
   o Yes
   o No

4. What is the proportion (%) of costs for administrative (e.g., security clearance for personnel, permits, etc), scanning, post-processing data, and modeling for:
   o Exterior
   o Interior – offices
   o Interior – mechanical/utility
   o Historic
   o Sculpture

5. How close are the actual costs to the original bids? The difference between original and actual bids is:
   - ○ < 10 %
   - ○ 10 % to 20%
   - ○ 26 % to 40%
   - ○ > 40 %

6. Are there instruments that are better suited for certain projects?
   - ○ Yes
   - ○ No

7. How often (% of the total number of projects you have worked on) is it discovered that during the modeling phase, more scans are needed or scans had to be repeated?
   - ○ Almost always (> 80 % of the total number of projects you have worked on)
   - ○ Often (> 50 %)
   - ○ Rarely (< 20 %)
   - ○ Never (0 %)

8. How much of a problem is interoperability between software packages
   - ○ Large problem
   - ○ Medium problem
   - ○ Small problem
   - ○ No problem

9. How do you suggest that GSA check the deliverables for:
   - ○ point clouds
   - ○ 2D plans
   - ○ 3D models

10. Excluding costs, what factors should GSA consider when evaluating bids?

11. Rank the following topics in order of discussion time spent during the workshop. 1 = needs most discussion to 10 (or 11 if Other is specified) = least discussion.
    - ○ Scope and specifications
    - ○ Estimate
    - ○ QAQC
    - ○ Best practices
    - ○ 3D Imaging lessons learned
    - ○ Rules of thumb
    - ○ Post-processing workflow
    - ○ Metrics
    - ○ Deliverables

- Future uses
- Other (Specify)

Appendix D

This appendix duplicates the comments received on the GSA's SOW for 3D-4D-BIM services document. This document was developed by GSA to provide guidance and sample language when preparing SOWs for 3D-4D-BIM services. The comments were received after the workshop was conducted and will therefore be in green text.

D.1 Contractor 1

Sample Scope of Services/Work Language

Page 14 – sentence This BIM Scope of Work will be a Firm Fixed Price Contract.
Comment: Consider removing as this statement should be included with the RFP and award documents.

Comment: In some instances it may be more appropriate to award task order as Time & Materials Contract

Page 14, Section 1.0 Objectives
Comment: In general, all narratives are appropriate and useful. However, ensure editor understands that rarely all will apply, and to utilize only those that meet specific project needs. Previous scopes of services received/reviewed by CoignAMT and partner firms have included several that did not apply.

Page 15, Section 2.0 3D Geometric Model
Comment: Should be included only when appropriate for project needs. Previous scopes of services received/reviewed by CoignAMT and partner firms have included this section and were not relevant to the project objectives.

Comment: It may be desirable to replace with Point Cloud based visualization tools such as fly-through animations, elevations depicting color or intensity, or cross-sections.

Page 17, Table 1
Comment: In general the table is well organized as a template. However, GSA PM may want to consider eliminating rows that are not relevant for a specific project as inclusion may lead to confusion.

Comment: Level 1 resolution does not match BIM Guide

Comment: Level 4 from BIM Guide not included

Comment: When both 2D and 3D deliverables are required, GSA may wish to reorder the options listing the 3D deliverable first. This will make it clear that the expectation is to extract 2D from the 3D/BIM model (thus saving time and reducing overall cost). Additionally, this approach will ensure the 2D and 3D deliverables are of the same detail and quality.

Comment: Uses of the terms "Tolerance" and "Artifact Size" was cause for significant discussion during the workshop. Many contractors preferred the idea of a list of components to include within the resulting model. However, CoignAMT is of the opinion that developing a list is not practical, and may not provide ample content. GSA may consider using both a Required Component List (i.e. pumps, valves, duct work, etc.) as well as the artifact size.

Comment: As one intended use of the point cloud dataset is to develop a repository of historical/as-is conditions, it may beneficial to include a minimum density (or spacing) for point clouds, and a second requirement for the modeling efforts. This approach would provide a very detailed point cloud, where artifacts are easily identifiable, yet provide a model that meets immediate needs and is cost appropriate for those uses.

Page 19: Specification of Deliverables

Comment: We believe the concept of a scan plan is very important to the success of the project. However, the scope of services requires the inclusion of equipment to be used, instrument calibration standards and current instrument calibration certificates. In our experiences, this information is not useful or understood by the GSA project managers. Additionally, as most contracts have access to multiple scanners, there is a possibility that this information provided within the Scan Plan, may not be for the actual unit deployed to the field. For that reason, the equipment and calibration information may be requested, but should be confirmed once field activities begin.

Comment: The 2nd paragraph of this section includes the statement *"the contractor shall specify whether the models to be submitted will be 3D geometric models or BIM"*. Consider removal of this statement. As this scope of work may be provide to multiple contractors for bid, the flexibility in the deliverables will lead to varied cost responses as developing a BIM model from a point cloud is much more time consuming than a 3D model.

Comment: The 3rd paragraph, while important, may lead to project delays and could have significant impact on "fast-tracked" projects. GSA should have the opportunity to review the Scan Plan and Post-Processing Plan, however approval and/or comments need to be provided to the contract within a short period of time.

Page 20: Registered Raw Data

Comment: The 1st paragraph defines ASCII as the preferred format for point cloud deliverables. While this is possible, ASCII formats are not efficient and result in much larger datasets.

Page 20: 2D Drawings of Existing Conditions

Comment: 2nd paragraph includes statement "the contractor shall extrapolate the drafted drawing beyond the limits of the laser scanner range to maximum extent possible showing as much detail as possible". This statement should be removed and be replaced with a statement or diagram that clearly defines the limits of the area of interest. The contract should take all steps necessary to provide data within this defined area and not beyond.

Comment: 3rd paragraph requires the submission of 2 sets of large paper drawings. As an alternative to paper deliverables, consider an all electronic submission to include PDF documents formatted to meet the 36" x 24" sheet size. This approach would control costs passed onto GSA, while allowing GSA personnel to produce hardcopy/paper products in the future if needed.

Page 22: Building Information Model

Comment: Consider consolidating this information with that included section 5.0 (page 23)

Page 24: "Note: This analysis is required"

Comment: Unclear if this note applies to section 5.1.1. or the following sections.

Comment: Section 5.2.1 – 5.6 include a series of components/features that are highlighted in yellow. Throughout the document, "highlighted" text indicates "optional or to be negotiated" items. Is the intent for the Scope Editor to develop a custom list of components for a specific project? If this is the case, consider developing a comprehensive list (or check list) of components to include as an appendix.

Comment: Sections 5.2.1 – 5.6, avoid use of "etc." in lists.

Page 30: 7.4 Performance Requirements/As-Built Requirements

Comment: This section is very brief, with the intent being unclear. Consider expanded.

Page 31: 8.2 Quality Control Report

Comment: This section is also too brief. Define specifically what GSA expects in terms of the report. Most information currently provides relates to bi-weekly status reports, not the QA report.

Page 34: 9.3 Status Report
Comment: This information is included within section 8.2 (page 31/32). Consider consolidating.

Page 34: 9.4 Post Scan Support Services
Comment: As opposed to defining a set period for support (6 month), consider including a number of hours. Current approach is very difficult for a contract to associate costs.

Page 35: 10.1 Cost
Comment: Consider utilizing both Firm Fixed Price and Time & Materials contracts. Some effort may include significant risk for the contractor, and prevent them from submitting a bid.

Page 35: 10.2 Personal and Past Performance and 10.3 Technical
Comment: Both sections seem to be written for non-IDIQ contract holders.

Page 36: 11.2 Site Visit and 11.3 Pre-Bid Conference
Comment: Both sections are important, however neither activity actually occurs.

D.2 Contractor 2

Executive Summary

Page 2:
Comment: Scenarios are confusing, and C is really like B with a focus on choosing applications by "phase". I'd replace pages 6 – 11 with the "charts" discussed in the NIST meeting.

Sample Scope of Services/Work Language

Pages 14-15: 1.0 Objectives
Comment: Focus on generating these with less focus on scanning and more on end use.

Pages 15-16: 2.0 3D Geometric model
Comment: Should be incorporated into objectives. Specifications to come from BIM SOW.

Page 17: Table 1
Comment: Replace (Table 1) by Scope (uniformity). Breakout. (with min. sys. sizes e.g., piping 1" or larger …)

Page 20: 1st paragraph

Comment: Use ASTM E57 format.

Comment: 1st sentence – data not raw anymore.

Pages 20 – 21: 2D Drawings of Existing Conditions, Plans, Elevations, Sections

Comment: Delete these sections. PBS CAD standards are expensive to provide in a scanning workflow. Requires re-drawing of 100 % of scanned scope. We suggest allowing simplified deliverables, digital only deliverables, etc.

Page 21: 3D/BIM Model of Existing Conditions

Comment: 1st paragraph. Remove the word "BIM" from section title, i.e., 3D Model of Existing Conditions. This section should only address 3D models. BIMs are specified in Section 5.

Comment: 2nd paragraph. Tolerances for modeling are impossible to deliver at scan tolerances. Modeling introduces additional error/tolerance and must be treated accordingly.

Page 22: Building Information Model

Comment: Remove this section – BIMs specified in Section 5.0.

D.3 Contractor 3

Page 21: 3D/BIM Model of Existing Conditions

Comment: Remove the word "BIM" from section title, i.e., 3D Model of Existing Conditions. 3D model of existing condition - Autocad/Microstation 3D) No Revit.

Page 21: Architectural

Comment: "The 3D model(s) shall include slabs/floors, walls ... ." Add columns to list.

Page 23: 5.0 Building Information Model

Comment: 2nd paragraph. Include survey drawings to "(Architectural and Structural)"

Page 33: Additional Sample Table of Deliverables

Comment: For Format = Interoperable Model, add ".rvt if agreed upon at start" to the Example File Types.

D.4 Contractor 4

Overall comments:

- SOWs should be built from successful projects

- From those projects, we create boiler plate sections to incorporate project types

- We need to keep the entire process simple. PMs are intimidated by the technology and projects.

- Identify other acquisition technologies for special circumstances – statues, etc.

Sample Scope of Services/Work Language

Page 14: 2nd paragraph. Firm Fixed Price Contract
Comment: Perhaps for the next 12 months, we should do cost plus to reduce risk.

Page 18: Referring to all tolerances listed on page and pg. 20 and 21
Comment: All resolutions and coverages directly affect the price of the project. With relaxed coverage requirement (80 %), the hours and pricing reduce.

Page 20: 1st paragraph
Comment: I would suggest removing "raw" from the deliverable. There is little to be gained until processing and cleaning.

D.5 Contractor 5

1. There are some general organizational comments about the SOW document that I feel need to be stated.

   a. The scenarios are confusing, and not particularly helpful in the new format we're proposing. It is not useful to define scanning requirements from "currently available technologies" any more than it is to make other project decisions based on technology. Determining requirements by phase is just a way of limiting use cases to a smaller subset, so it is really Scenario B anyway. Personally, I'd like to see this section disappear.

   b. 2D and 3D deliverables should get their own section (instead of being part of section 3) as they are unrelated to data capture or processing. I could even see section 3 broken into sections highlighting capture and processing requirements separately. As much as possible, the document's organization should mirror the process both the GSA and the contractors will be going through. BIM sections should be referenced to the BIM SOW, and as those relate to laser scanning that should be covered by the modeling LOD (level of detail) requirements for as-built and as-is modeling in the BIM SOW not the Laser Scanning SOW. This simplifies the Laser Scanning SOW and reflects that often times modeling is simply unnecessary.

c. Examples worked into the sample scope of work are useful only if they are clearly marked as examples. Sometimes, it is difficult to tell what is an example and what is recommended practice. Either a formatting or organizational change needs to occur to make this exceedingly clear.

d. Future use is a challenge for scanning projects today. I know we made this point several times in the meeting, but I didn't see it represented as clearly as I'd like in the NIST report. I would however point something out in hindsight. As long as the scanning (today) is done with clearly defined coverage, tolerance, and resolution then it is very simple to incorporate these current scans with future data thus allowing the current scans to have an extended life without requiring 100% data capture or having to go back and re-capture 100% of the data later on. This might guide us to define standards that can reduce the current cost of scanning without precluding what you all are wanting to do long term.

2. I wanted to re-emphasize the suggestion of moving to the E57 format as a standard. While still in draft status, I think this format is what the industry needs and the weight of the GSA behind it would push the straggling companies (trimble, leica, etc...) to support it.

3. There is a serious gap in information in all three documents regarding error/tolerance/accuracy/precision/whatever you want to call it. Aside from picking and sticking with a term and definition to refer to this as throughout the documentation, there is a need to bring it back in line with reality. Scanning is an abstraction or reality with certain measurement and registration tolerances. I feel like the documents overall have a handle on this and describe it fairly well (although in some cases you reference scanning to 1/8" tolerance when most lasers are producing a spot larger than 1/8" which pretty much eliminates the ability to provide anything to 1/8" tolerance given edge conditions and sub-pixel errors). However, right next to this the documents request an identical tolerance for modeled deliverables (referenced to the "true" measurement). This is simply impossible to provide. Modeling or drafting is a further abstraction from the point cloud and introduces its own errors and deviations. The document either needs to reflect this by adding the tolerance of modeling and the tolerance of measurement together (and therefore defining the modeling tolerance separately) or it needs to define modeling tolerance as being measured from the point cloud and not the "true" measurement. Either way is fine, although I think the former is less confusing. With an 1/8" tolerance on the measurement and an 1/8" tolerance on the modeling the best you can ask for is a model that is 1/4" +/- the "true" condition.

4. Construction tolerances are well defined (though hard to find) for different systems. Frequently, defining a consistent tolerance (1/8" for example) is unnecessary given that some construction solutions incorporate a much larger tolerance in their assembly. Curtain wall connections for instance usually allow +/- 1" in the horizontal and +/- 2" vertically from their connection points, if not more. Locating an imbed to +/- 1/8" is just a waste of money in that scenario. This same issue propagates across the whole scope matrix. These industry standard tolerances need to be understood when defining scope requirements for accuracy, and lends more weight to the practice of defining scanning requirements by scope.

5. I strongly believe that the GSA needs to move to a project wide Scope/Phase based LOD matrix. This is what we're doing on all our projects, and it works very well. In short, we sit down with any currently contracted project stakeholders and define the acceptable LOD for each scope by project phase. This is a living document that can be revised as new team members are contracted. However, it gives us a single document to reference all our BIM, Coordination, Laser Scanning, and other contract exhibits to. Thus, we don't re-define these scopes multiple times on a project. Once for BIM, once for Coordination, once for Laser Scanning. all to end up with different requirements and some being insufficient for use in other deliverables. This would also simplify the Laser Scanning and BIM SOW documents as the LOD sections can be all but removed and referenced to the project documentation. Then, you can work on a training program to help your PMs fill out the LOD document for projects (require one for all projects perhaps?) and these will naturally fit into the other referencing documents. The exact format is something we're still working on, but I'd be happy to share a current version.

www.ingramcontent.com/pod-product-compliance
Lightning Source LLC
Chambersburg PA
CBHW081904170526
45167CB00007B/3139